Igor Marques-Carneiro

Remote Control System

Igor Marques-Carneiro

Remote Control System

of Telecommunications Equipment for Contingency

ScienciaScripts

Imprint

Any brand names and product names mentioned in this book are subject to trademark, brand or patent protection and are trademarks or registered trademarks of their respective holders. The use of brand names, product names, common names, trade names, product descriptions etc. even without a particular marking in this work is in no way to be construed to mean that such names may be regarded as unrestricted in respect of trademark and brand protection legislation and could thus be used by anyone.

Cover image: www.ingimage.com

This book is a translation from the original published under ISBN 978-613-9-62248-1.

Publisher:
Sciencia Scripts
is a trademark of
Dodo Books Indian Ocean Ltd. and OmniScriptum S.R.L publishing group

120 High Road, East Finchley, London, N2 9ED, United Kingdom
Str. Armeneasca 28/1, office 1, Chisinau MD-2012, Republic of Moldova, Europe
Printed at: see last page
ISBN: 978-620-7-71925-9

SUMMARY

DEDICATORY

I dedicate this work, in particular, to my family, especially my wife Barbara Cristina Chaves de França Marques Carneiro, and my daughter Alice Chaves de França Marques Carneiro, who is yet to come into this world, for all their support during the time I was developing this project.

Finally, to everyone who believed and believes in my potential.

INTRODUCTION

The online world is already a reality. Every day, new applications and equipment are launched on the market that can be controlled or monitored via the Internet. We can list various household appliances such as fridges, washing machines and even coffee makers that already have versions with integrated Internet access. In addition to this equipment, we can also easily find other highly access-dependent equipment in homes, such as IP monitoring cameras, alarms, VoIP telephones[1] and, above all, Internet routers. On the other hand, telecommunications companies are expanding their networks in terms of capacity and capillarity in order to offer services with greater capacity to as many subscribers as possible. These expansions have led to an increase in the number of small remote centers known as PoPs (Point of Presence [1]), which house equipment that controls and distributes access to a given area or even serves as a transit point for other PoPs. Some of these PoPs even house dozens of pieces of equipment such as routers, switches, optical modems, radios and other telecommunications equipment. Although the standard for this telecommunications equipment is stricter than for domestic equipment, often some of this equipment can suffer a "crash", losing all control over the network. In these cases, the problem can lead to a PoP outage affecting all subscribers in the region. Most of these crashes could be quickly resolved with a simple restart of the equipment (commonly known as a reboot), carried out by a technician at the location where the equipment is located. In cases where a PoP can serve as a transit point for other PoPs[1] , any outage can have a cascading effect on thousands of subscribers.

This growing dependence on the Internet network for home or residential use is also generating a growing demand for quality and, above all, availability of access from the operators offering the service. So any failure of a PoP can lead to a veritable avalanche of complaints at the operators' call centers and even sanctions from their regulatory agencies. Therefore, the speed with which access is recovered becomes an important indicator of the quality of the services provided to subscribers, as well as reducing the number of complaints to be dealt with by the operators.

CHAPTER 1 OBJECTIVE

The motivation for this work was the development and implementation of a device capable of controlling the power supply to the equipment of a broadband Internet and VoIP telephony operator in its PoPs. One of the main problems encountered by this operator was the delay in recovering its PoPs and even customer access, due to the size of its network.

We can see the typical access topology of this operator in Figure 1.1. Currently, the urban commute in certain parts of the city turns a journey of a few kilometers into a journey of many hours. In addition to this commute, some PoPs are located on the roofs of commercial and residential buildings, making access difficult and even restrictive at certain times. Therefore, in the event of a failure, the service may be partially or totally inoperable for many hours until the team is able to reach and access the site. In some cases, the recovery time can be more than 6 hours, causing a large number of complaints and even cancellations. As previously mentioned, in the case of PoPs where there is relaying to other PoPs, the problem can affect thousands of subscribers. Most of these events used to occur simply due to an equipment outage where a simple reboot would temporarily solve the problem until the problem was solved definitively or the equipment was replaced. The use of IP-controlled sockets such as the "IP Power Controller 9258T Aruca Electronics, [2]" has access to the operator's own network link, which would often not solve the problem, as the outage could affect the network's edge equipment in the PoP, thus isolating the recovery device itself, preventing any remote reboot. So we thought of a device that had independent access to the operator's network that would perform a remote reboot until the technician arrived on site. Knowing that in these PoPs the only Internet network available was that of the operator itself, we thought of a device that would use the cell phone network to control the power supply to the telecommunications equipment. As a result of this need, the system described below was devised.

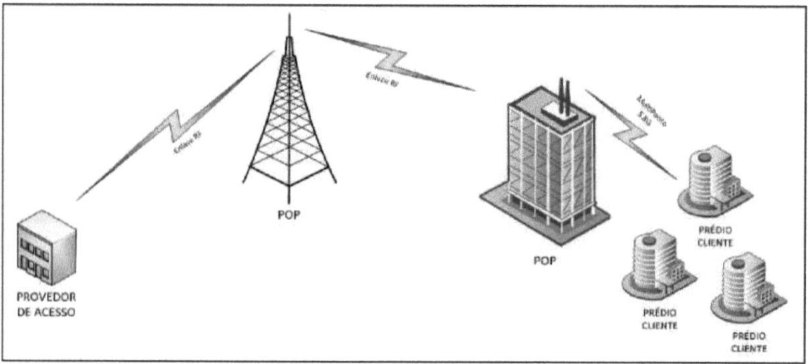

Figure 1.1- Typical operator topology

CHAPTER 2 DEVELOPMENT

This section describes the technical details used to develop this project. The initial access base was a simple cell phone, which could be a used handset with a self-service function. From this point on, access to the circuit was considered via the handset's Headset interface (headphones combined with a microphone). The circuit was designed to recognize DTMF tones [3] (*DualTone Multi-Frequency*) in the audio signal transmitted through the call automatically answered by the mobile handset. Thus, from any call made to the cell phone attached to the circuit, it was possible to transmit commands to the circuit.

2.1 SISTEMADTMF

According to Wikipedia 2013, DTMF signaling was developed at Bell Labs to enable DDD dialing, which uses wireless links such as microwave and satellite.

DTMF stands for "Dual-Tone Multi-Frequency", the two-frequency tones used to dial the most modern telephones (Wikipedia, 2013);

The combination of frequencies that make up DTMF tones, shown in table 2.1, where the intersection of a high frequency, described in the first row, and a low frequency, described in the first column on the left. The intersection of the row and the column, i.e. the intersection of the two frequencies, forms the respective DTMF tone.

Hz	1209	1336	1477	1633
697	1	2	3	A
770	4	5	6	B
852	7	8	9	C
941	*	0	#	D

Table 2.1 - DTMF table (Source: Wikipedia, 2013)

To detect the DTMF tone, the MT8870DE IC [4] shown in figure 2.1 was used. This IC has the characteristic of detecting and decoding all 16 pairs of DTMF tones and placing a 4-bit code on its output terminals (Q4,Q3,Q2,Q1) (shown in table 2.2),

According to its datasheet:

The MT8870D/MT8870D-1 monolithic DTMF receiver offers small size, low power consumption and high performance. Its architecture consists of a bandsplit filter section, which separates the high and low group tones, followed by a digital counting section which verifies the frequency and duration of the received tones before passing the corresponding code to the output bus. (p.4).

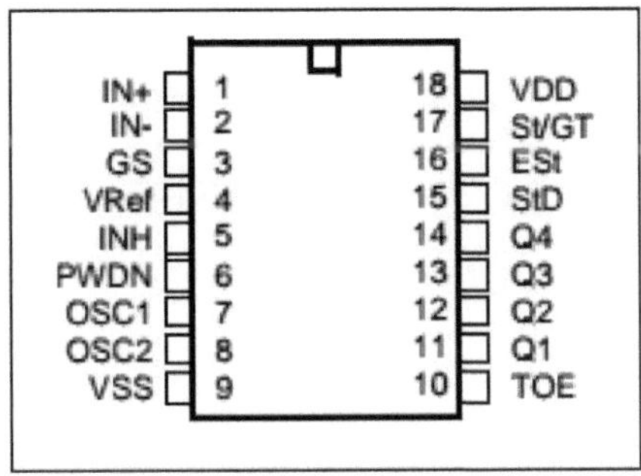

Figure 2.1 - MT8870DE pinout (Source: MT8870DE datasheet)

Type	Q4	Q3	Q2	Q1
1	0	0	0	1
2	0	0	1	0
3	0	0	1	1
4	0	1	0	0
5	0	1	0	1
6	0	1	1	0
7	0	1	1	1
8	1	0	0	0
9	1	0	0	1
O	1	0	1	0
*	1	0	1	1
#	1	1	0	0

Table 2.2 - Binary terminal balance table (Source: MT8870DE datasheet)

2.2 MICROCONTROLLER

The microcontroller used was the ATmegal68 [5] (figure 2.2 and figure 2.3) manufactured by the Atmel Corporation. Although it is a microcontroller with less presence on the market than PIC-based controllers, the ATmegal68 has the advantage of being the same one used by the Arduino project [6], an easy-to-use programming, compiling and recording tool with extensive documentation and examples. This made it possible to use all the development tools in Arduino in the initial phase of developing the project's embedded software. The main features of the ATmegal68 microcontroller are:

- I6k of flash memory;

- 5I2 Bytes of EEPROM;

- Ik Byte of RAM;

- two 8bit counters and one I6bit counter;

- 6 PWM (Pulse-width modulation) channels;

- 8 I0-bit ADC channels in TQFP andMLF

- 6-channel I0-bit ADC inPDIP;

- 23 programmable input and output ports;

- operating voltage 2.7 to 5.5V;

- processor: 0- I2 MHz @ 2.7 - 5.5V, 0-24 MHz @ 4.5 - 5.5V;

- low consumption: I MHz, I.8V: 240µA, 32 kHz, I.8V: I5µA;

Figure 2.2 - ATmegal68 (Source: HWTech, 2008)

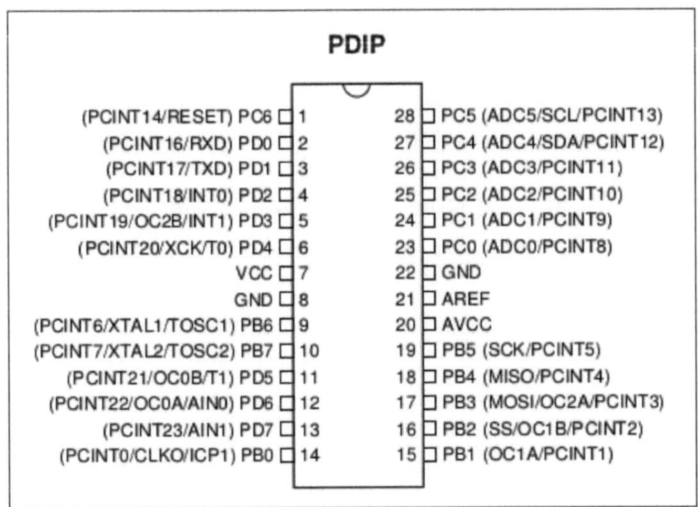

Figure 2.3 - ATmegal68 pinout (Source: ATmegal68 datasheet)

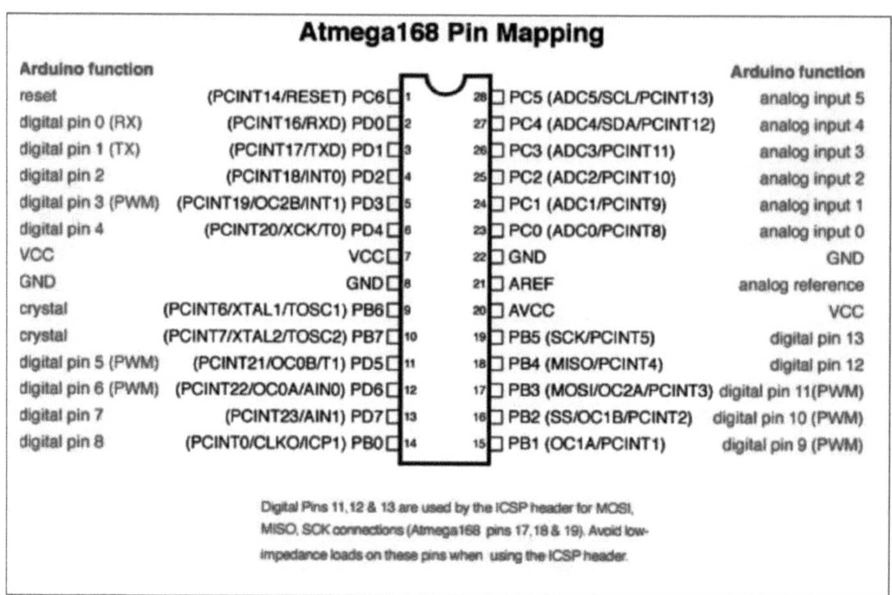

Figura 2.2   - Pinout definition according to arduino (Source: http://arduino.cc)

2.3 CONTROL SOFTWARE

As mentioned earlier, Arduino was used to develop the software due to the ease of the programming interface (Figure 2.5). In turn, it was necessary to write the Arduino bootloader to the microcontroller. The language used to develop the embedded software was C, which can be found in full in APPENDIX A, which was then compiled by the Arduino IDE into assembly instructions and then written directly to the microcontroller.

Basically, the software works in an infinite loop, i.e. at the end of the instruction it returns to the beginning by consulting the information inputs, when it finds an expected input, the instruction is routed to the next function. The software checks two types of input, one is the input on pin 25, which in the code is defined by the name "A2" (see figure 2.3), this input is connected to a pushbutton[2] which, when pressed, gives a high signal (5 volts), at which

point, in the software, the intrusion recognizes this signal, waits 10 seconds

and checks again if the signal is still high, i.e. the button is still pressed, the

process of resetting to the default password begins, which in this case has

been set to the password "1234", directly in the EEPROM memory and then

returns an audio signal which will be explained below. The other check is the

input on pin 15, which in the code is called port "9". This input is connected to

the CI MT8870DE and receives a high signal when a 4-bit code is available

for reading. At this point, the software reads the 4 bits on ports Q1, Q2, Q3

and Q4, then converts the bits into decimal using the formula in table 2.3 and

calls the "create_phrase()" function, passing as a parameter the number

obtained which refers to the DTMF digit. This function first returns an audio

signal with a short tone indicating that the digit has been received. This lets

us know if the digit has been recognized or even if a duplicate has been

received, which is very common in cases of echo[3] in telephony. The function

then checks which digit was keyed and adds it to an external variable[4] to form

the command. Digit 12, which is the "#" key, indicates that this is the end of

the command, thus calling the "trigger_command()" function. The purpose of

this function is to analyze the contents of the external variable, which we call

"phrase", where a certain number of digits have been saved, and to execute

different processes for each condition of this phrase, as described in table

---

2 "[...]such as press, depress, mash, and punch[...]" http://en.wikipedia.org/wiki/Push-button
3 It's a reflection of a sound that returns to its source http://en.wikipedia.org/wiki/Echo
4 Variable created outside the context of the function and will be recognized by other functions.

2.4.

$$
(bit04 * 8) + (bit03 * 4) + (bit02 * 2) + (bit01 * 1);
$$

Table 2.3 - binary to decimal conversion

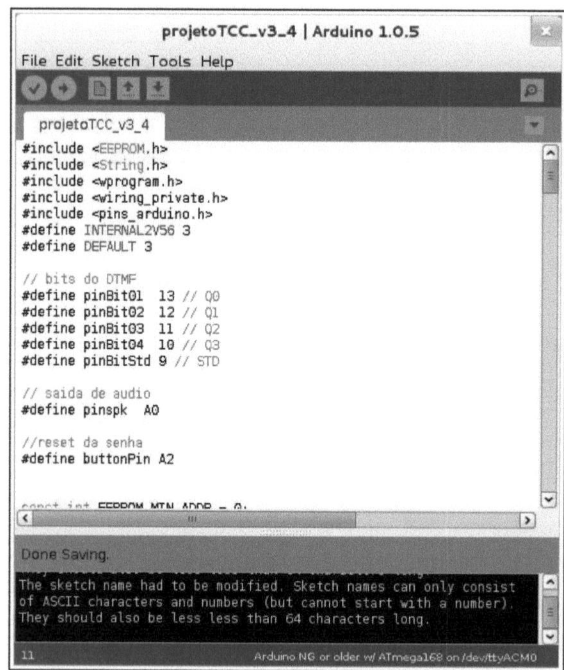

Figure 2.5 - Arduino development software screen

command to change the password:

sentence content: <current password>*<new password>*<new password again>#

example: 1234*9876*9876#

command to turn off the tap:

sentence content: <password>*<plug number>*0#

example: 1234*1*0#

command to switch on the socket:

sentence content: <password>*<plug number>*1#

example: 1234*1*1#

reboot command and plug in a socket:

sentence content: <password>*<plug number>*3#

example: 1234*1*3#

command to check the current status of the socket:

sentence content: *<plug number>#

example: *3#

Table 2.4 - list of commands

The commands to change the password, turn the socket off, turn the socket on and reboot[4] , after being executed, return a long, high-pitched tone in audio form to inform you that the command was successful. The command that checks the current status of the socket returns a long, high-pitched tone to inform you that the socket is in the current on state or a long, low-pitched tone to inform you that the socket is in the current off state. After each command is executed, the content of the sentence is deleted to save the next entry.

Eight LTV-4N25 opto-couplers [7] manufactured by Fairchild Semiconductor were used for the power driver part to drive the relays in the socket strip. The Tornece optocoupler provides convenient electrical isolation, as well as making it possible to use an external voltage source in its second stage. In this way, we can use this device simply to short the 8 output terminals, enabling the relay to be powered by another voltage source, as well as simply closing a circuit with the device's power supply.

The optocoupler works very simply: an infrared diode is connected to terminals 1 and 2 and when activated by a voltage of around 1.5v, the diode starts emitting infrared, thus activating the photosensitive transistor at terminals 4 and 5, as shown in figure 2.6.

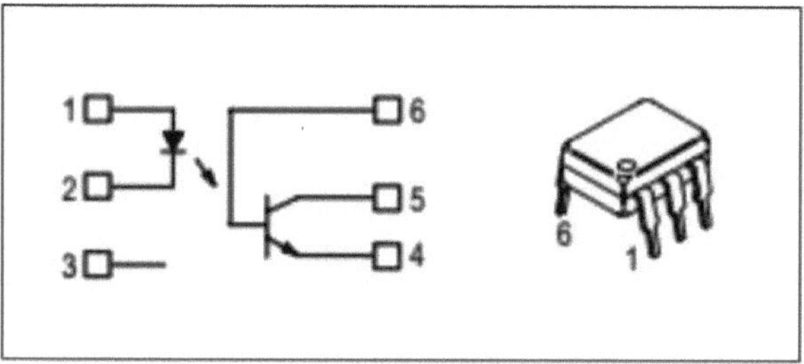

Figure 2.6 - Schematic of the LTV-4☐25 optocoupler (Source: Datasheet)

2.5 TAKEOVERS

To control the sockets, we used eight standard 5V, 127V IOA relays connected as normally closed and a 1N4148 diode connected in parallel to the relay coil, inside the structure of the ruler. This solution proved to be the most practical because the control circuit could be changed and maintained without disconnecting the equipment connected to it.

According to Newton C Braga [8]:

[...] The purpose of the diode is to protect the circuit against an inverse voltage generated when the relay is de-energized and a voltage is induced. This voltage has the opposite polarity to that which created the field and can reach very high values. It can cause serious damage to the circuit.[...] (2013)

The socket strip was connected to the circuit using a multi-way cable with a DB9 connector. The switch box made it possible to install the relays internally without the need for mechanical adaptations, allowing the relays to be placed close to the sockets. Another advantage of the switchgear chosen was that it was 19" standard, allowing it to be used in a rack, a standard widely used in telecommunications. Figure 2.7.

Figure 2.7 - Socket strip (Source: brdshop.com.br)

2.6 FINAL CIRCUIT

The final circuit was developed using the "CadSoft EAGLE [9]" software and is shown in figure 2.8. It shows the presence of an LED connected to port 5 on the MT8870DE IC, whose purpose is to signal the recognition of the

DTMF digit, facilitating any tests on the equipment. There is also a trimpot[5] connected to pins 2 and 3 of the MT8870DE IC, which is responsible for fine-tuning the gain of the IC's audio input. As the circuit depends on the use of a cell phone, we have provided two types of power supply for the cell phone, a USB plug type power supply due to the widespread use of this type of connection to recharge the battery, and a power supply via a PCF1-02 type connector, both with the circuit's own 5V power supply. To activate the relays, which are inside the sockets, two configurations have been provided, controlled by two jumpers arranged in parallel and with two positions on each one, as shown in Figure 2.9. For the relays to use the circuit's own power supply, i.e. the 5V supply, simply connect the jumpers in the "INT" position. Otherwise, setting the jumpers to "EXT" will completely isolate the circuit from the connections to the relays by isolating the optocoupler. In this way, it will be necessary for the socket strip to have its own power supply for the relays, making it possible to install devices with a greater load.

---

5 is a miniature adjustable potentiometer http://pt.wikipedia.org/wiki/Trimpot

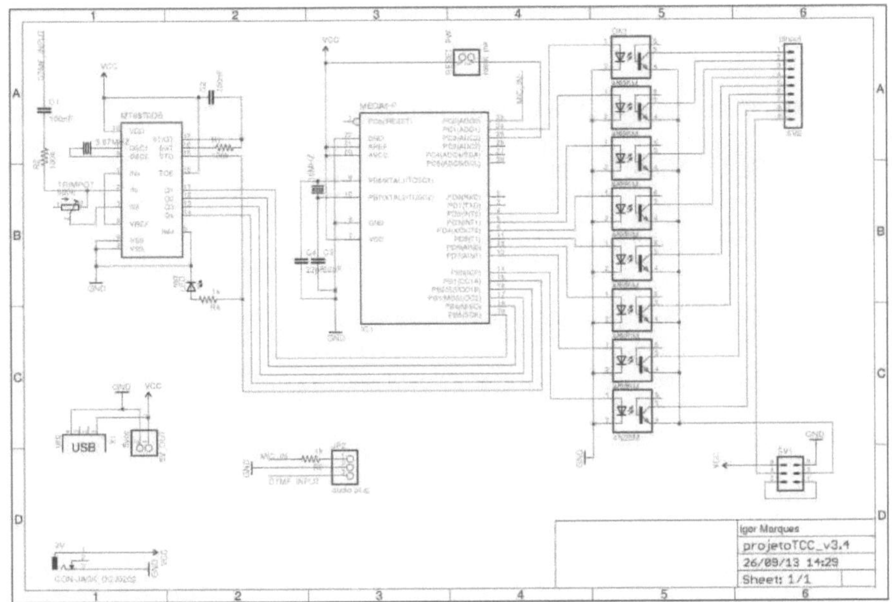

Figure 2.8 - *Circuit diagram*

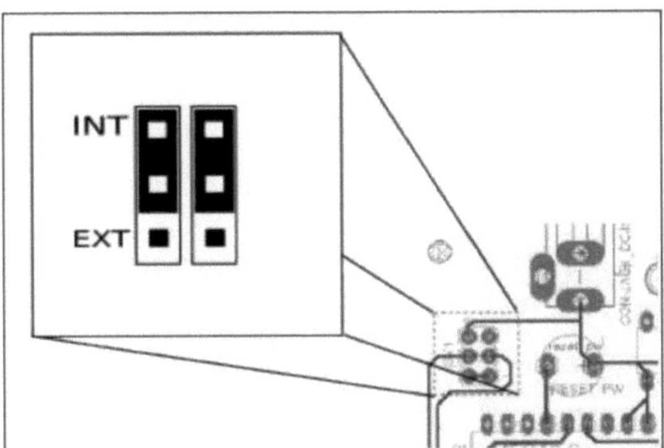

Figure 2.9 - Relay power selection switch

CHAPTER 3 IMPLEMENTATION

Using the circuits described above, a printed circuit board was also designed using the "CadSoft EAGLE" software (shown in Figure 3.1).

According to the website www.esquemas.org [10]:

Physically, we can describe printed circuit boards in general as follows: An insulating substrate, being o phenolite or fiber, with a thickness of approximately 1.6 mm; and a very thin conductive copper laminate surface, approximately 0.05 mm, glued to the insulating base by industrial processes[...] (2013)

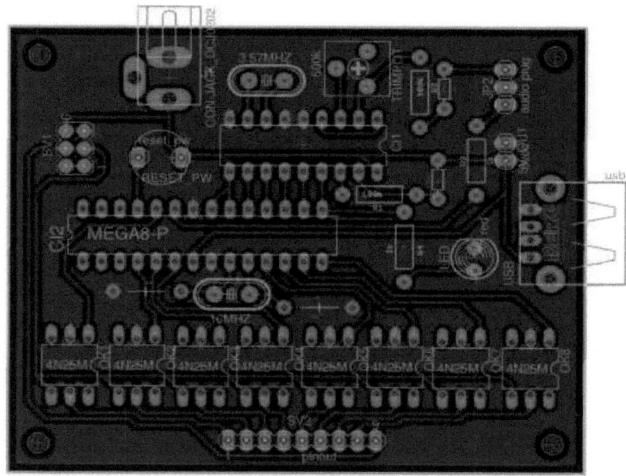

Figure 3.1 - Final circuit obtained using Eagle

3.I TRANSFERRING THE IMAGE TO THE PLATE

Three ways of transferring the circuit to the board were tested. Transfer by laser printing on photographic paper, transferred by applying heat to the board. This method proved to be simple but with very poor resolution. The second method tested was the use of DrY-Film, (a photo film sensitive to ultraviolet light) this method proved to have the highest quality in terms of resolution, but the disadvantage of this method is that DrY-Film is sold in large quantities and costs as much as R$ 690.00 for a box with two rolls 20cm wide and 153 meters long (Dupiza, 2013). Lastly, and the one chosen for the development of this project, the screen printing method was tested, which consists of transferring the image printed on transparent paper to a screen or commonly called SilkScreen through a photographic process (figure 3.2). 180-strand nylon was used for this, as it is an image with a lot of detail in the order of 0.1mm.

Figure 3.2 - Ready-made silk screen

For this project, "Fiberglass" board was chosen. Although its cost is high compared to phenolite board, its quality is superior in terms of mechanical resistance and adhesion of the copper to the board. To transfer the image, which had already been developed on the silk screen, we used GNOOI-I ink from the manufacturer Gènesis, which is resistant to corrosion. Placing the screen on the fiberglass mat, which should already be cut to size, we applied the ink on top of the screen and then spread it with a squeegee suitable for Silk Screen. At the end of this process, we get the fiberglass board printed with the circuit design as shown in figure 3.3. After this process, although the board has the circuit design, it still has copper lamination all over it, so the next corrosion step is necessary.

Figure 3.3 - Board with o circuit design

3.2 PLATE CORROSION

The part where the copper on the fiberglass board is exposed, i.e. not covered by the GNOOI-I paint, must be removed so that only the copper covered by the paint remains, thus forming the final circuit. To remove the copper, there are various corrosion methods such as the use of nitric acid (HNO3), generally used in high-speed industrial production processes. For artisanal processes, iron perchloride (FeCl3) or ammonium persulphate (NH$_4$) can be used. In the case of artisanal production, the corrosion time depends on the area and thickness of the metal to be corroded, as well as the multiple reuses of the solution. On average, it takes around IO minutes to corrode a plate measuring IO x IO cm (Wikipedia, 2OI3). For this project, we used iron perchloride as shown in figure 3.4. After corrosion (figure 3.5), it was necessary to remove the paint that remained on the copper tracks. To do this, we submerged the board in sodium hydroxide "NaOH" and with the help of a small brush the paint came off easily, revealing the copper track that was under the paint.

Figure 3.4 - Plate corrosion

Figure 3.5 - Plate after corrosion

3.3 WELDING MASK APPLICATION

Solder mask, also known as solder resist, is a thin layer of special paint that resists the high temperature when soldering to prevent the solder from touching other vias or the ground of the circuit when the components are being soldered. The purpose of the solder mask is also to protect the copper, which is exposed to prevent corrosion in the heat.

There are various types of solder mask, such as the DrY-Film type mentioned earlier, which also comes in the form of a solder mask. For this project, we used the same screen-printing process that we used to transfer the circuit design to the board. The only difference in this step is that the ink used must be different because the ink's required resistance is no longer corrosion but heat and weather resistance. For this purpose, we used T2011-1 paint, also from the manufacturer Gènesis. You can see in figure 3.2 that we used the same silk screen that we used for the image transfer process to house the solder mask, which we then transferred to the circuit board using the silkscreen process, where you can see the board with the solder mask in figure 3.6.

Figure 3.6 - Plate with welding mask

3.4 PLATE DRILLING

To drill the holes where the components will be attached, we used a mini drill suitable for jobs requiring greater precision. The holes were drilled in two sizes: for components with thicker pins, such as the CI terminals, we used a 0.8mm drill bit, and for the other holes, for components with smaller pins, such as capacitors and resistors, we used a 0.6mm drill bit.

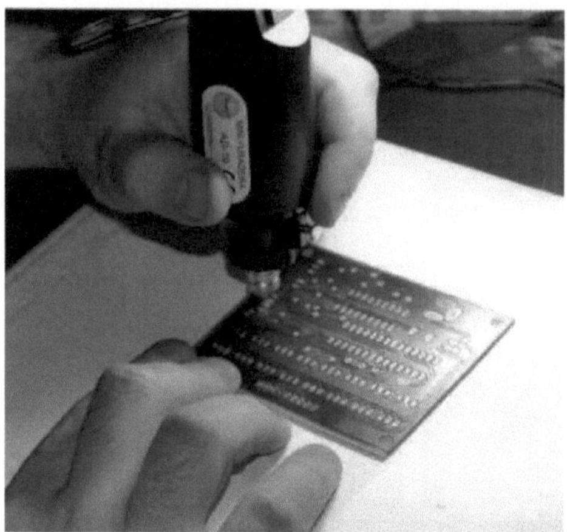

Figure3.7- Drilling the plate

3.5 FINAL ASSEMBLY OF THE SYSTEM

Each component was soldered to the board using a 60-watt soldering iron. In order not to damage the MT8870DE and ATmegal68 CI's, we included a socket for each CI in the project so that, as well as not damaging them during soldering, they could be removed later for reprogramming, in the case of the ATmegal68 CI. Figure 3.8 shows the finished board with all the components.

To protect the board, we put it inside a plastic box commonly called a patola box, which is often used for prototype development. You can see in figure 3.9 that the box used is much larger than the board, so there is enough space left over to fit a used cell phone.

For the connection to the multi-way cable, we prefer to install a DB9 type connector, due to its high mechanical resistance, attached only to the box and connected with parallel wires to the board, so that the board does not suffer from possible forces applied when connecting the external multi-way cable.

Figure 3.8 - finished board

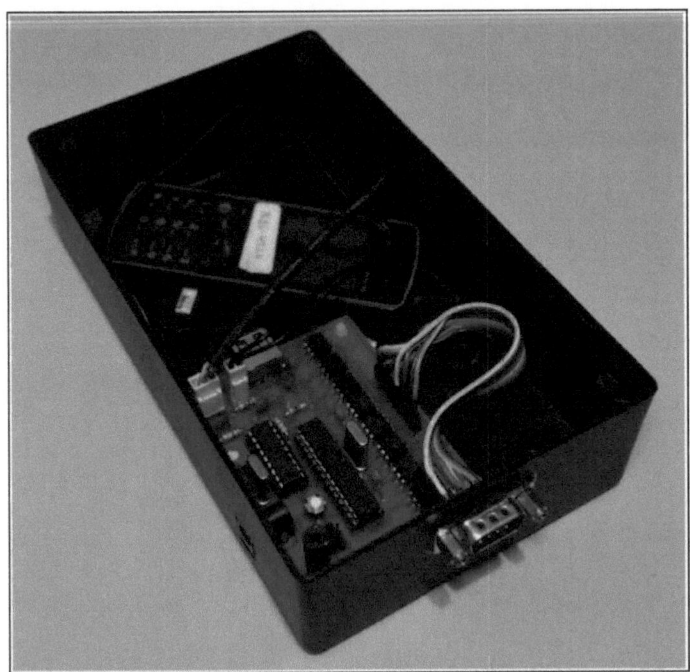

Figure 3.9 - final assembly

3.6 PRACTICAL APPLICATION

Pilot tests were carried out at MLS Wireless S/A, a provider of broadband Internet access and Voip telephony in the state of Rio de Janeiro. With the authorization and support of the current president Rogério Passy[1] , tests were carried out with the device proposed in this project at a relay point in Barra da Tijuca in the west of Rio de Janeiro, a place most affected by the delay in recovering access due to its distance from the company's headquarters in the Botafogo district in the south of Rio de Janeiro. At this relay point, better known internally as POP, the ruler was installed in a rack connecting eight pieces of equipment, which are RF link equipment such as PowerBridge, NanoBridge and airFiber, all from the manufacturer UbiQUiTi. A Cisco-SF300 switch and an optical modem were also connected. Figure 3.10 shows the installation in the 19" rack where the box containing the system proposed in this project is installed above the eight socket strip.

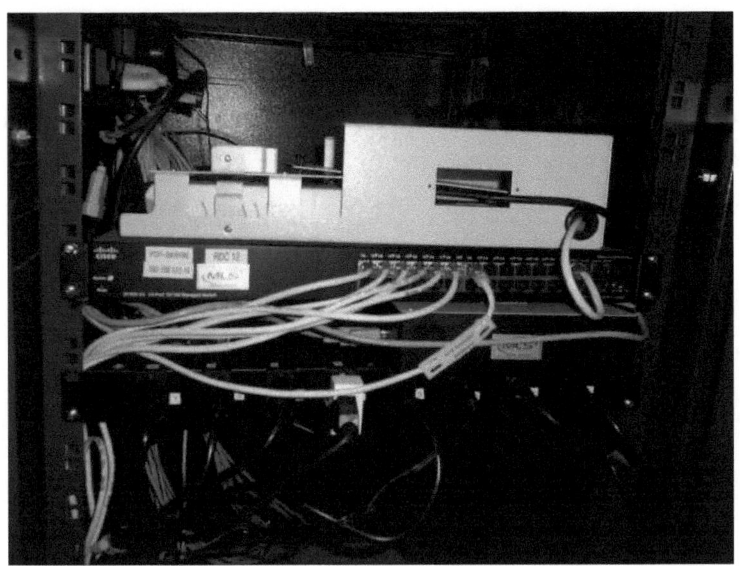

Figure 3.10 - POP-BARRA06 installation

3.7 PROJECT COST

N	material	quantity	unit value
1	Crystal 3.57MHZ	1	R$ 2,00
2	Crystal 16MHZ	1	R$ 3,00
3	100nF capacitor	2	R$ 0,10
4	22pF capacitor	2	R$ 0,10
5	CM8870CP	1	R$ 5,00
6	Atmega168	1	R$ 11,00
7	DC Power Jack 2mm	1	R$ 1,50
8	Polarized connectors 3	1	R$ 1,00
9	Polarized connectors 2	1	R$ 1,00

10	Red LED	1	R$ 0,10
11	4N25 optocoupler	8	R$ 1,50
12	100k resistor	2	R$ 0,10
13	1k resistor	2	R$ 0,10
14	Push down button	1	R$ 1,50
15	500k trimpot	1	R$ 1,00
16	USB Female 90º	1	R$ 2,50
17	Female pin bar connector 9	1	R$ 1,50
18	2x3 male pin bar connector	1	R$ 1,50
19	Socket 18 pins	1	R$ 1,50
20	24-pin socket	1	R$ 2,00
21	Fiberglass board 20x30 cm	1/9	R$ 30,00
total		**32**	**R$ 52,23**

Table 3.1 - Construction per plate

N	material	quantity	unit value
1	Paiola box	1	R$ 5,00
2	DB9 male connector	1	R$ 1,00
3	Relay 127VAC 5VDC	8	R$ 1,50
4	Diode 1N4148	8	R$ 0,15
total		**18**	**R$ 19,20**

Table 3.2 - Other costs

N	type	UN.	Qty	unit	total
1	Hh development cost	Hh	100	R$ 50,00	R$ 5000,00
2	Implementation cost ~ 10 pieces	Hh	5	R$ 10,00	R$ 50,00
3	ART costs	1	1	R$ 60,00	R$ 60,00

Table 3.3 - Hh and ART costs

CHAPTER 4 SUSTAINABILITY

Have you ever met someone who needed to throw away a useless cell phone and didn't know where to dispose of it? Around 40 million tons of electronic waste are generated every year in the world. Among emerging countries, Brazil is the country that generates the most e-waste.

Research by the Santo André Foundation reveals that 4,770 tons of cell phones, including batteries and chargers, will be discarded in landfills across the country this year. In 2013, the amount will reach 7,500 tons[...].

According to environmental engineer Nathâlia de Carvalho Aiolfi, who carried out the study, 48 million devices with an average weight of 100 grams were sold in the country in 2010. "Their lifespan is between two and three years and today it's very difficult to dispose of this material. Most of it goes in the general waste," he says. In the state of São Paulo, 136 cell phones are registered for every 100 inhabitants. In October 2011, 57,000 handsets were sold, an increase of 11.2% on the same period in 2010.

With the National Solid Waste Policy, instituted by Law No. 12.305 of August 2, 2010, which came into force last year, manufacturers, importers, distributors and retailers are responsible for the life cycle of products. The legislation also creates obligations for public urban cleaning agencies and consumers. All are subject to fines for non-compliance. The fines range from R$ 500 to R$ 10 million.

The new law obliges industries and their entire supply chain to manage these materials, adopting measures for their correct disposal and implementing selective collection.

Consumers are also obliged to comply with the legislation by returning their electronic waste to the industry.

Composition of cell phones [11]

45% plastic;

10% ceramics;

20% copper;

20% gold, aluminum and other metals;

5% non-metals. [9]

Problems caused by improper disposal [12]

- This is done when the equipment becomes defective or obsolete (outdated). The problem occurs when this material is disposed of in the environment. As this equipment contains chemical substances (lead, cadmium, mercury, beryllium, etc.) in its composition, it can cause soil and water contamination.

- In addition to polluting the environment, these chemicals can cause serious illnesses in people who collect products from rubbish dumps, wastelands or the street.

- This equipment is also made up of a large amount of plastic, metals and glass. These materials take a long time to decompose in the soil. [10]

In this project, it was possible to reuse outdated cell phones that the company would have had to discard in order to provide the means of communication for the circuit, contributing to the reduction of electronic waste caused by the rapid technological evolution of cell phone handsets. A survey carried out at MLS Wireless, where the prototype was tested, showed that on average the company's group cell phone contract is renewed every two years and with each renewal, the operator offers a number of new cell phones in the package to replace the employees' older handsets.

CONCLUSION

The solution presented proved to be more efficient, as it offers alternative access to the operator's network available on site. In addition, the cost of the solution represents a fraction of the equipment available on the market. The ruler's independence also proved to be efficient in cases where the control circuit is changed or maintained, increasing the uptime of the equipment controlled by the system. The use of a used cell phone not only reduces the cost of a possible GSM module to be used, but also contributes to the reduction of electronic waste caused by the disposal of outdated cell phones.

In future implementations of the project, we will be able to increase the number of control ports to cover a greater number of sockets to be controlled, as well as inserting the functionality to check whether a socket is actually in use by simply checking the current flow. We could also develop a mobile app for Android and iOS to register the numbers and define buttons referenced to each socket so that when the user clicks this button, the app will automatically send DTMF commands to the predetermined equipment. In order to make the product better prepared for the market, in the future we could implement a voltage regulator circuit at the input so that it is better isolated from input voltage variations or even the use of out-of-specification sources. We could also implement a microSD recording circuit to record operation logs. We could also insert an "ISD1420" type CI to store audio files to provide as feedback instead of sound signals.

BIBLIOGRAPHICAL REFERENCES

[1]   ,   Point   of   presence,,   Available   at:
<http://en.wikipedia.org/wiki/Internet_service_provider>, Accessed on: July 2, 2013...

[2]   Aruca Electronics, IP Power Controller 9258T, Available at:
<http://www.amazon.com/IP-   Power-9258T-Network-
Controller/dp/B005D55HKU>, Accessed on: Jul. 2013.

[3]   WIKIPEDIA, the free encyclopedia, DTMF,, Available at:
<http://pt.wikipedia.org/w/index.php?title=DTMF&oldid=34485343>, Accessed on:
Jul. 2013.

[4]   MITEL, Integrated DTMF Receiver, May 1995.

[5]   Atmel, Atmel Microcontroller, Apr. 2011.

[6]   Arduino S/A, Arduino,, Available at: <http://www.arduino.cc>, Accessed on:
Aug. 2012.

[7]   LITEON, General Purpose TypePhotocoupler,,Sept. 2013.

[8]   According to the Newton C Braga Institute,,, Available at:
<http://www.newtoncbraga.com.br>, Accessed on: Jul. 2012.

[9]   CadSoft Computer GmbH and CadSoft Inc, CadSoft EAGLE, Available at:
<http://www.cadsoftusa.com>, Accessed on: Feb. 2012.

[10]   Esquemas.org,, Available at: <http://www.esquemas.org>, Accessed on: Feb.
2013.

[11]   TIM CELULAR S.A, Composiçâo dos aparelhos celulares,, Available at:
<http://www.tim.com.br>, Accessed on: Nov. 2013.

[12]   UNEP - UN Environment Program, Problems caused by improper disposal, Available at: <http://www.suapesquisa.com/o_que_e/lixo_eletronico.htm>, Accessed on: Nov. 2013.

APPENDIX A PROGRAM SOURCE CODE

```cpp
#include <EEPROM.h>
#include <String.h>
#include <wprogram.h>
#include <wiring_private.h>
#include <pins_arduino.h>

#define INTERNAL2V56 3
#define DEFAULT 3
#define pinBit01  13 // Q0
#define pinBit02  12 // Q1
#define pinBit03  11 // Q2
#define pinBit04  10 // Q3
#define pinBitStd 9 // STD

// saida de audio
#define pinspk  A0
//reset da senha
#define buttonPin A2
const int EEPROM_MIN_ADDR = 0;
const int EEPROM_MAX_ADDR = 511;
const int BUFSIZE = 5;
char buf[BUFSIZE];
String myString;
char myStringChar[BUFSIZE];
String frase = "";
int bit01 = 0;
int bit02 = 0;
int bit03 = 0;
int bit04 = 0;
int tecla = 0;
// portas das tomadas
int tomadas[]= {0,A1,2,3,4,5,6,7,8};

// checagem se a tomada esta desligada ou ligada
int checar_on_off(uint8_t pin){

    uint8_t bit = digitalPinToBitMask(pin);
    uint8_t port = digitalPinToPort(pin);
    if (port == NOT_A_PIN) return LOW;
    return (*portOutputRegister(port) & bit) ? HIGH : LOW;
}


boolean eeprom_is_addr_ok(int addr) {
        return ((addr >= EEPROM_MIN_ADDR) && (addr <= EEPROM_MAX_ADDR));
}
boolean eeprom_write_bytes(int startAddr, const byte* array, int numBytes) {
        int i;
        if (!eeprom_is_addr_ok(startAddr) || !eeprom_is_addr_ok(startAddr + numBytes))
                return false;
        for (i = 0; i < numBytes; i++)
                EEPROM.write(startAddr + i, array[i]);
        return true;
```

```
}

boolean eeprom_write_string(int addr, const char* string) {
        int numBytes;
        numBytes = strlen(string) + 1;
        return eeprom_write_bytes(addr, (const byte*)string, numBytes);
}


boolean eeprom_read_string(int addr, char* buffer, int bufSize) {
        byte ch;
        int bytesRead;
        if (!eeprom_is_addr_ok(addr)) {
                return false;
        }
        if (bufSize == 0) {
          return false;
        }
        if (bufSize == 1) {
          buffer[0] = 0;
          return true;
        }
        bytesRead = 0;
        ch = EEPROM.read(addr + bytesRead);
        buffer[bytesRead] = ch;
        bytesRead++;
        while ( (ch != 0x00) && (bytesRead < bufSize) && ((addr + bytesRead) <=
EEPROM_MAX_ADDR) ) {
                ch = EEPROM.read(addr + bytesRead);
                buffer[bytesRead] = ch;
                bytesRead++;
        }
        if ((ch != 0x00) && (bytesRead >= 1)) {
                buffer[bytesRead - 1] = 0;
        }
        return true;
}
void comando_no(){
        tone(pinspk, 200,1000/4);
        delay(1000/2);
        noTone(pinspk);
        tone(pinspk, 200,1000/4);
}
void comando_yes(){
        tone(pinspk, 200,1000);
}
void criar_frase(int digito){
        if(digito == 12){
                dispara_comando();
                frase = "";
        }else if(digito == 11){
                frase += "*";
                tone(pinspk, 200,1000/7);
        }else if(digito == 10){
                frase += "0";
```

```cpp
                tone(pinspk, 200,1000/7);
        }else{
                frase += String(digito);
                tone(pinspk, 200,1000/7);
        }
}
void dispara_comando(){
        char senha_salva[BUFSIZE];
        Serial.println("comando: ");
        Serial.print(frase);
    //consultar senha na EEPROM
    eeprom_read_string(0, senha_salva, BUFSIZE);
  if(String(senha_salva) == frase.substring(0,4)){
        // senha digitada confere
    if(frase.length() == 14){
     //procedimento de troca de senha
            if(frase.substring(5,9) == frase.substring(10,14)){
                // senhas iguais pode salvar
                myString = frase.substring(5,9);
                myString.toCharArray(myStringChar, BUFSIZE);
                strcpy(buf, myStringChar);
                eeprom_write_string(0, buf);
                comando_yes();
            }
    }else if(frase.length() == 8){
    //comando das tomadas
        if( (frase.substring(5,6).toInt() >= 1  ) && ( frase.substring(5,6).toInt() <= 8 ) ){
                if(frase.substring(7,8) == "0"){
                        /***** desliga tomada *****/
                        digitalWrite(tomadas[frase.substring(5,6).toInt()], HIGH);
                        comando_yes();
                }else if(frase.substring(7,8) == "1"){
                        /***** liga tomada *****/
                        digitalWrite(tomadas[frase.substring(5,6).toInt()], LOW);
                        comando_yes();
                }else if(frase.substring(7,8) == "3"){
                        /***** desliga tomada por 3 s depois liga *****/
                        digitalWrite(tomadas[frase.substring(5,6).toInt()], HIGH);
                        delay(3000);
                        digitalWrite(tomadas[frase.substring(5,6).toInt()], LOW);
                        comando_yes();
                }
        }
    }
  }else if(frase.length() == 2){
        if(frase.substring(0,1) == "*"){
        /*** comando para checar se a tomada esta ligada ou desligada ***/
            if( (frase.substring(1,2).toInt() > 0  ) && ( frase.substring(1,2).toInt() < 9 ) ){
                if(checar_on_off(tomadas[frase.substring(1,2).toInt()]) == 0){
                        // tomada ligada
                        tone(pinspk, 800,1000);
                }else{
                        // tomada desligada
                        tone(pinspk, 100,1000);
                }
```

```
                    }
                }
            }
}
void setup() {

        Serial.begin(9600);
        analogReference(DEFAULT);
        pinMode(buttonPin, INPUT);
        analogWrite(buttonPin, LOW);
        pinMode(pinspk, OUTPUT);
        noTone(pinspk);
        // colocando estado inicial para todas ligadas
        pinMode(A1, OUTPUT);
        digitalWrite(A1, LOW);
        pinMode(2, OUTPUT);
        digitalWrite(2, LOW);
        pinMode(3, OUTPUT);
        digitalWrite(3, LOW);
        pinMode(4, OUTPUT);
        digitalWrite(4, LOW);
        pinMode(5, OUTPUT);
        digitalWrite(5, LOW);
        pinMode(6, OUTPUT);
        digitalWrite(6, LOW);
        pinMode(7, OUTPUT);
        digitalWrite(7, LOW);
        pinMode(8, OUTPUT);
        digitalWrite(8, LOW);
}
void loop(){
 if(analogRead(buttonPin) > 1000 ){
    delay (10 * 1000);
    if(analogRead(buttonPin) > 1000 ){
        myString = "1234";
        myString.toCharArray(myStringChar, BUFSIZE);
        strcpy(buf, myStringChar);
        eeprom_write_string(0, buf);
        comando_no();
    }
}

 if (digitalRead(pinBitStd) == HIGH) {
  bit01 = digitalRead(pinBit01);
  bit02 = digitalRead(pinBit02);
  bit03 = digitalRead(pinBit03);
  bit04 = digitalRead(pinBit04);
  tecla = (bit04 * 8) + (bit03 * 4) + (bit02 * 2) + (bit01 * 1);
  criar_frase(tecla);
  delay (500);
 }
}
```

Printed by Books on Demand GmbH, Norderstedt / Germany